ANCIENT GIANTS OF THE FOREST

Contents

Chapter 1

Remarkable Trees

Many people have a favorite tree. Maybe there is an oak tree in your yard that you love to climb. Maybe there is an old maple tree that you love to sit under on a hot summer day. Or maybe you have visited an orchard where you can pick peaches to eat.

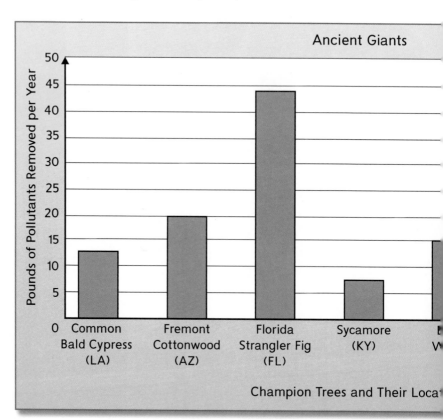

Champion Trees and Their Locat

This bar graph shows how many pounds of pollutants trees remove from the air every year. (*Champion trees* are the largest-known type of a specific tree.)

Trees help people in many ways. Trees help clean the air. Trees are places for animals to hide or live. Trees make the world more beautiful.

tree
Poplar
)

Rocky Mountain
Ponderosa Pine
(MT)

Chapter 2

How Do You Measure a Champion?

What makes a tree a champion? It could be the oldest, largest, tallest, or strongest tree. Some trees have beautiful fall colors. Other trees have survived a hurricane or a drought.

This bristlecone pine tree, is more than 4,700 years old. Can you think of anything else that is as old as this tree?

Average crown spread

Girth

Copyright © McGraw-Hill Education (bkgd) ©Royalty-Free/Corbis; (r) Jamie And Judy Wild/Danita Delimont, Agent/Alamy

A group called American Forests decided how to compare trees. The **girth**, height, and **crown** of a tree are measured. Then the numbers are used to get a total point value. The value is used to name a new champion tree.

Did You Know?

In 50 years, one tree recycles more than $37,000 worth of water, provides $31,000 worth of erosion control, $62,000 worth of air pollution control, and produces $37,000 worth of oxygen.

A tree's girth is the circumference, or the distance around the tree. The girth is measured $4\frac{1}{2}$ feet above the base, or bottom, of the tree.

A tree gets one point per inch of circumference. A tree that has a girth of 40 feet would get 480 points.

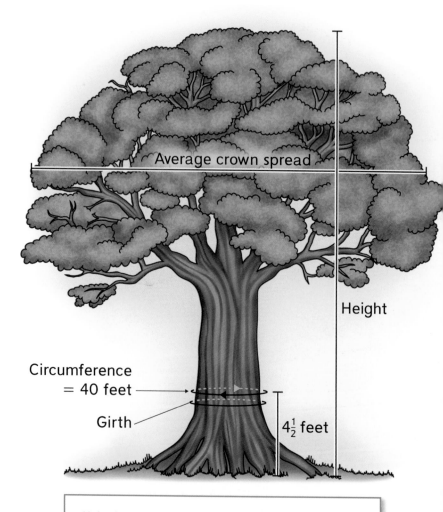

Average crown spread

Height

Circumference
= 40 feet

Girth

$4\frac{1}{2}$ feet

12 inches = 1 foot. A circumference of 40 feet would measure 40 × 12, or 480 inches around.

One tree has the record for largest girth. This tree is a giant sequoia in Sequoia National Park. This tree is called General Sherman. It measures 85 feet around. People think General Sherman could be the largest living thing in the world!

The General Sherman tree was named after the famous Civil War general—General Sherman. What is the total number of girth points for this giant Sequoia?

The second part of deciding total point value is height. A tree is measured from the base of the trunk to the very top branch.

This Texas live oak measures 42 feet from bottom to top.

Height might seem like an easy measurement to make on a short tree. But some trees can be hundreds of feet tall.

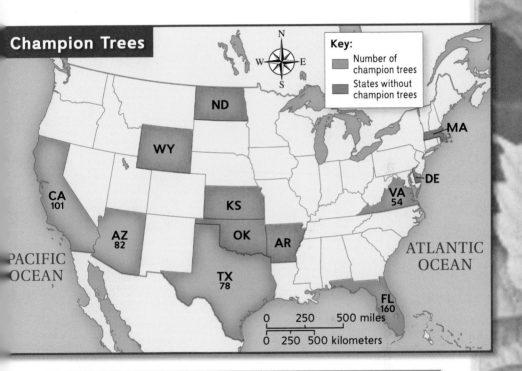

Champion Trees

Key:
Number of champion trees
States without champion trees

ND

MA

WY

DE

CA
101

KS

VA
54

AZ
82

OK

AR

ATLANTIC
OCEAN

PACIFIC
OCEAN

TX
78

FL
160

0   250   500 miles

0   250   500 kilometers

The 5 states shown in green have the greatest number of champion trees. The 7 states shown in red do not have any champion trees.

The tallest national champion is a coast redwood tree at Jedediah Smith Redwoods State Park in California. The tree measures 321 feet tall. This redwood is about as tall as a 26-story building!

How long does it take to build a building? How long do you think it would take for a redwood to grow to the same height?

Some champion trees are more than 2,000 years old.

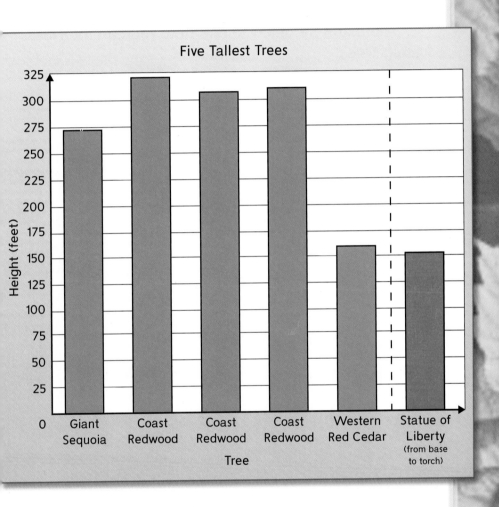

Five Tallest Trees

Height (feet): 25, 50, 75, 100, 125, 150, 175, 200, 225, 250, 275, 300, 325

Tree: Giant Sequoia, Coast Redwood, Coast Redwood, Coast Redwood, Western Red Cedar, Statue of Liberty (from base to torch)

Math DETECTIVE

How could you tell whether 252 feet is a good estimate of the giant Sequoia's height? Provide a good estimate of the coast redwood's height.

People who look for big trees use a math problem to estimate the height of a tree. If a tree could be on the list of champions, then an expert is asked to make exact measurements.

The example below shows how you can measure a big tree.

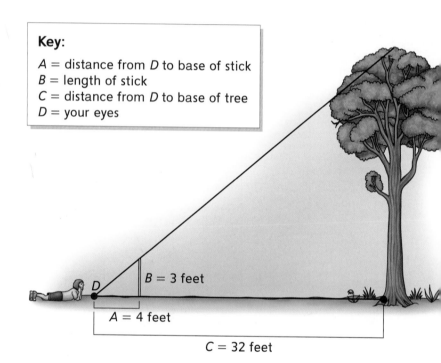

Key:

A = distance from D to base of stick
B = length of stick
C = distance from D to base of tree
D = your eyes

B = 3 feet

A = 4 feet

C = 32 feet

$$\frac{B \times C}{A} = \text{height of tree} \qquad \frac{3 \times 32}{4} = 24 \text{ feet}$$

Lying on the ground (D), place the stick (B) so that the top of it lines up with the top of the tree. Ask a friend to make the measurements for A, B, and C.

There are tools that can measure large objects. A clinometer, a relascope, or a laser can help someone get measurements.

A clinometer measures angles. A laser is often used with a clinometer to measure a tree's height.

The last part of measuring a big tree is the crown. The crown is the leafy part of a tree. The crown is measured at the narrowest point and widest point. These two numbers are added. The sum is divided by two. The answer is the average width of the tree's crown.

91 feet

35 feet

To find the crown of this tree, add the two measurements (35 + 91 = 126) and divide by 2. (126 ÷ 2 = 63)

An equation can be used to figure out a tree's total points:

$$\text{girth} + \text{height} + \frac{\text{crown spread}}{2} = \text{total points}$$

Each tree is compared to other trees of the same kind. The tree with the greatest number of points for a species is the champion.

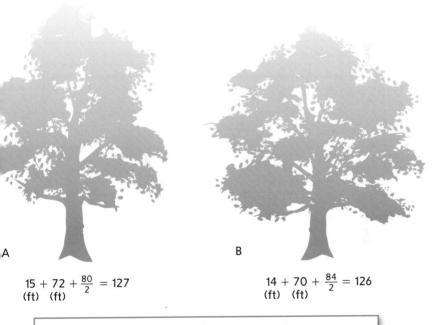

A

$$15 + 72 + \frac{80}{2} = 127$$
(ft) (ft)

B

$$14 + 70 + \frac{84}{2} = 126$$
(ft) (ft)

Look at the equations. You can see that Tree A has more total points than Tree B. Tree A would become the champion for its kind.

Chapter 3

List of Champions

The National Register of Big Trees has been checking trees since 1940. Each year, this group checks the list. They remove the trees that have died. The Register adds the names of the new champions.

Champion Trees				
Tree	State	Girth (inches)	Height (feet)	Average Spread
Giant Sequoia	CA	1020	274	10
Coast Redwood*	CA	950	321	7!
Coast Redwood*	CA	895	307	8:
Coast Redwood*	CA	867	311	10
Western Red Cedar	WA	761	159	4!
Sitka Spruce*	WA	668	191	9(
Sitka Spruce*	OR	629	204	9:
Coast Douglas Fir	CA	512	301	6!

* Shows co-champions

There are three trees that have been on the list since t was created. They are the giant sequoia (in Sequoia National Park), the Rocky Mountain juniper (in Utah's Cache National Forest), and the western juniper (in Stanislaus National Forest in California).

This champion western juniper was discovered by shepherds during the 1920s. It stands 78 feet tall. It is more than 4,000 years old.

Older trees have a better chance of being a champion. Older trees they have had more time to grow. One way to estimate the age of a tree is to count the number of rings in a tree's cross-section. But seeing a cross-section means cutting down the tree.

A different way to find a tree's age is to measure the girth of the tree about 5 feet above the ground. A tree with a tall, straight trunk, growing close to other trees might struggle for sunlight. So, every 5 inches of girth is equal to about one year of growth. (See *B*)

A tree might stand alone with plenty of side branches growing from the trunk. For this tree, every 1 inch of girth is equal to about one year of growth. (See *A*)

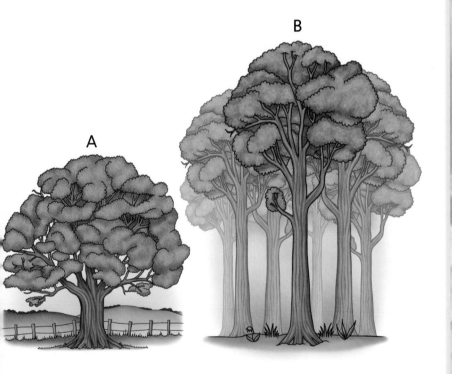

B

A

Chapter 4

Hunting for Trees

Some people enjoy a hobby called **geocaching** (JEE oh cash ing). They use a tool called a GPS to find sm hidden items. Geocachers start with numbers that tell t **longitude** and **latitude** of the item. Longitude and latitude numbers can help you find any place on Earth.

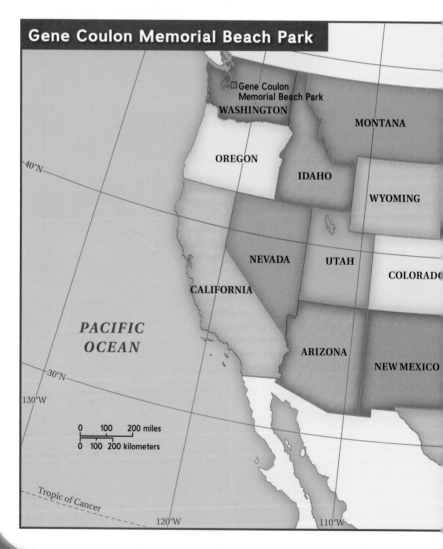

Gene Coulon Memorial Beach Park

Copyright © McGraw-Hill Education   (bkgd) ©Royalty-Free/Corbis

The National Register of Big Trees lists the longitude and latitude numbers of several trees on public land. Tree hunters can use a compass or GPS to find the champion trees.

Tree hunters can find the champion sitka willow at Gene Coulon Memorial Beach Park using these coordinates: 48° N latitude, 122° W longitude.

If you want to be a tree hunter, there are many big trees to find. There are 826 kinds of trees in the United States.

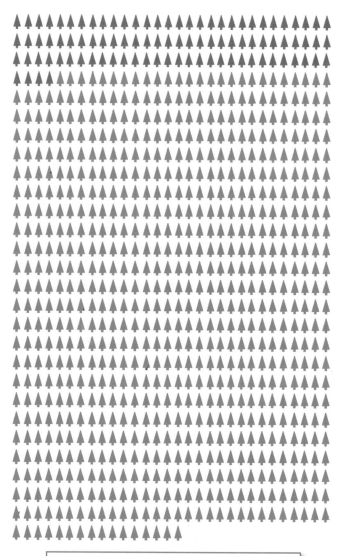

Each tree above stands for a kind of tree. The green trees do not have national champions.

Only 94 of the 826 species have national champions. Maybe you could find one. What better way to help save these big trees?

Glossary

crown

> The upper part of a tree, consisting of foliage and branches. *(page 5)*

geocaching

> A sport or hobby involving hiding and finding objects, using GPS (global positioning system) data to log the location of the cache and share this data with potential players. *(page 20)*

girth

> The distance around a thick cylindrical object such as a tree. *(page 5)*

latitude

> Distance north or south of the Equator; measured in degrees. *(page 20)*

longitude

> Distance east or west of the Prime Meridian (an imaginary line at 0° longitude); measured in degrees. *(page 20)*